AMERICAN FIRE APPARATUS CO.

1922-1993 PHOTO ARCHIVE

Richard J. Gergel

Iconografix

Photo Archive Series

Iconografix
PO Box 446
Hudson, Wisconsin 54016 USA

Library of Congress Control Number: 2004102902

ISBN 1-58388-131-X

04 05 06 07 08 09 6 5 4 3 2 1

Printed in China

Cover and book design by Dan Perry

Copyediting by Suzie Helberg

Cover photo- One of the many stock model American Fire Apparatus pumpers manufactured for quick delivery, this American/Ford C-850 had a 1,000-gpm pump, 500-gallon tank, 6-man canopy tilt cab, and rear personnel windshield. *Photo courtesy of Bob Hook*

BOOK PROPOSALS

Iconografix is a publishing company specializing in books for transportation enthusiasts. We publish in a number of different areas, including Automobiles, Auto Racing, Buses, Construction Equipment, Emergency Equipment, Farming Equipment, Railroads & Trucks. The Iconografix imprint is constantly growing and expanding into new subject areas.

Authors, editors, and knowledgeable enthusiasts in the field of transportation history are invited to contact the Editorial Department at Iconografix, Inc., PO Box 446, Hudson, WI 54016.

Acknowledgments

Many thanks to the following whom without hesitation shared their extensive collections of fire apparatus photographs and literature to make this Photo Archive of American Fire Apparatus a reality. A special acknowledgement to Dick Adelman, Mark Boatwright, Dan Decher, William Friedrich, Warren Gleitsmann, Gary Handwerk, Steve Hagy, Bob Hook, Dennis Maag, Chuck Madderom, Larry Phillips, Dave Organ, Jerry Spotts, Greg Stapleton, and Dave Stewardson.

To the valued new and old friends the Author has been re-acquainted with or made new who contributed their individual private information: Henning O. Anderson, Dee Enos, Elaine Jewell, Carl Leyda, Betty and Keith (nee Bentley) Mead, and Bruce Upston.

Not to be forgotten are the many past employees and salesmen of American Fire Apparatus both at Battle Creek, Michigan and Marshalltown, Iowa who shared their knowledge of the events that occurred.

Last but not least, to the Fire Departments who have given technical information and photographs in an effort to memorialize the American Fire Apparatus Company.

Bibliography

Battle Creek Enquirer and News, Battle Creek, MI
Marshalltown Times-Republican newspaper, Marshalltown, IA
Marshalltown Public Library, Marshalltown, IA
Stratford Beacon Herald newspaper, Stratford, Ontario, Canada
Willard Library, Battle Creek, MI
Donald F. Wood & Wayne Sorensen, *Big City Fire Trucks, Volume 1 1900-1950*, Krause Publications
Walter McCall, *American Fire Engines Since 1900*, Crestline Publishing Co.
Bruce Upston, *American Fire Apparatus History*, High School Term Paper 1959

An aerial view of the American Fire Apparatus administrative and manufacturing facility (approx. 38,600 sq. ft.) in Battle Creek, Michigan. Since 1936 the company produced over 4,500 municipal pumpers and aerial ladders when Collins Industries acquired it on March 1, 1978. Most of the units were produced on a commercial chassis, the stronghold of the company's business for over 42 years. The Company produced a custom-built fire apparatus chassis (one of the heaviest in the industry), aerial water towers (the largest in the field), industrial foam trucks (one of the highest quality in the industry), aircraft crash fire rescue apparatus (a new concept in this complex field), and aerial platform equipment (very versatile and flexible). The main factory in Battle Creek, Michigan [top left] (1948-1978). Also shown is a manufacturing facility [top right] in Marshalltown, Iowa (1955-1976).

Introduction

Entrepreneurial businessmen Messer's Andrus, Beach, Burnham, Hyde, Nichols, Shepard and Ward established the company name, originally called "Machinery Equipment Company," in 1859 in Battle Creek, Michigan. In 1880 the name changed to American Steam Pump Company. Then in 1891, an engineer named Marsh designed and patented the first successful steam valve which was unique in that it required no linkage with a piston rod for operation. The company's name changed to the American Marsh Pump Company. In 1922 Ben D. Barton made a contract with the Company, "builders of centrifugal pumps since 1875," and with the help of their engineers developed a power take-off for driving centrifugal force pumps (an idea he had for mounting a pulley on the front of a Cle-Tractor engine to run belt-driven machinery) on tractors and eventually on passenger cars and light trucks. The idea developed over several years while he sold Caterpillar tractors for Cletrac Corporation. In 1924 Mr. Barton started to work for American Marsh on a full-time scale and eventually he became sales manager of their pump sales. He toured the country for American Marsh Pumps to sell their front mounted pumps. The pump could easily be installed on the popular (and inexpensive) Ford Model T truck to provide fire protection for small towns and rural areas. The pumps were sold under the name of Barton's Products Company until about 1930 when the name was changed to American Fire Apparatus and Equipment Company. During the Depression of the 1930s, Barton-equipped fire apparatus became even more popular as cash strapped departments sought to replace older equipment. In the early 1930s the fire industry desired a mid-ship pump design and the company developed an entirely new concept called the "Duplex Multi-stage" pump, unlike any others being offered in the marketplace. The new concept offered two separate impellers on opposing shafts. In 1936 American Marsh started to design and build the new concept mid-ship pump.

In 1937 American Fire Apparatus Company was formed. Mr. Barton had the first fire trucks built in Battle Creek in January 1937. Six were built by local body shops. Barton operated from his house with his wife as his secretary. She ordered accessory items, filled under-writers reports and performed many other duties. In late 1937 Barton made an agreement with Anderson Coach Company, manufacturers of house trailers, located 10 miles east of Battle Creek, to build its fire trucks. The first one of 30 trucks was built for Sipesville, Pennsylvania (Job No. 208). In November 1939 Mr. Barton rented an old brick building on Jackson Street and, along with three other men, began building fire apparatus in a 40-foot-square space plus about 40-square-feet of space for storage. The only equipment that they had consisted of a small lathe, a small power hacksaw, an arc welder, one acetylene welder and cutter attachment, several drill motors, a portable sander, and a few other small tools. Sketches were prepared and sent to a local steel supplier for cutting and bending the tank and body parts. The first truck built by American Fire Apparatus in its own plant was a Rural Community truck with a front mounted pump. It was stationed at Berne, Indiana (Job No. 236).

More men were hired in the winter of 1939-1940 and by midsummer 1940 most of the storage space was used for manufacturing space. At that time there were eight men on the payroll. The second floor was used for storage space. On December 7, 1940, it was announced that Ben D. Barton, Adrian P. Adney, and Harold Bentley were granted the patent for a duplex pump, adapted for mounting so that it could be operated by the truck engine, which allowed for two pumps of the rotary type to be driven and operated independently or, if high pressure was desired, a connection was provided for passing water through the two pumps in series. This patent was assigned to American-Marsh Pumps Inc. Many other products were patented during this time, which included the rear dump, the engine governor, and the vacuum primer as well as several other new models of pumps; single-, two-, three-, and four-stage pumps. The business grew and by the spring of 1941 an adjacent room was made into an office and an assistant was hired to assist Mrs. Barton. More men were hired in the winter of 1939-1940 and by midsummer of 1940 most of the storage space was used for manufacturing.

American-Marsh Pumps Inc. employed Henning J. Anderson, who immigrated to the United States from Sweden as a boy of nine in 1925, and progressed through the ranks to General Manager. Mr. H. J. Anderson became the President of American-Marsh Pumps in 1936.

American Fire Apparatus was sold to American-Marsh Pump Company. American Fire Apparatus retained its own name, continuing to build fire apparatus and using pumps by American-Marsh. Mrs. Barton became the office manager. Late in 1941 another factory was purchased, this one on Hall Street. This building was 60 feet wide by 200 feet long with an office space on the second floor. By this time, there were 20 workers in the factory and three in the office. In 1942 the company received a contract to mount 20 pumps for the Corps of Engineers on chassis furnished by the government. As World War II continued three additional contracts were received from the government calling for the building of 220 complete fire trucks with front mounted pumps. Many were shipped to Western Europe to protect the cities and battlefields. Because of this, no chrome was allowed to show. Many times various items that were manufactured before the War that were chrome plated had to be painted over with olive drab paint. They also built approximately 100 skid units for the Office of Civil Defense that could be loaded onto the backs of pick-ups or flatbed trucks for emergency use. Very few civilian trucks were built during the last 3-1/2 years of the War due to the shortage of steel. American Fire Apparatus decided to do something about it. They made several wood body trucks. At the end of World War II there was a big demand for fire equipment but new chassis were not readily available. Many communities furnished a used chassis and American Fire Apparatus made it into a shiny new fire truck. Business increased and in 1947 Mr. Barton converted an old shed on his small farm north of Battle Creek into a building suitable for building booster tanks, oak racks for use under the fire hose bed, and the machining and grinding of castings. The business grew and in July 1948 a new 42x150-foot cement block building, on four acres south of Battle Creek at the corner of Main and Golden Avenue, was occupied. This was the last location of the company until it was sold in 1978. Soon another 42x150-foot building was established. This included manufacturing space and a new office headquarters. It was occupied in May of 1950.

In short order, another 66x150-foot addition was built and, in February 1951, the factory downtown was closed and its operations moved into this latest addition. All operations were under one roof with the exception of a steel Quonset test house, which stood on the same property about 200 feet away from the main building on the bank of a creek. Here the finished fire trucks underwent several hours of testing and, occasionally, a six-hour test was made for the National Board of Underwriters.

American Fire Apparatus manufactured 250 to 300 units per year with a work force of 100 employees in Battle Creek, Michigan. Because of the sales volume being generated, the company continued to expand, establishing additional manufacturing plants, first in Stratford, Ontario, Canada (1951) to penetrate the Canadian market under the American-Marsh Pumps Canada, Ltd. label. The manufacturing facility (Approx. 7,000 square feet) was built at Monteith Avenue in Stratford, Ontario, Canada in early autumn 1951. An additional 5,000 square feet was later added to the plant. During their best years they produced 150 units per year and had approximately 40 employees.

In 1956 American Fire Apparatus Company was set up as a separate corporation with most of the stock owned by the stockholders of American-Marsh. The Company was credited with the first U. L.-approved front mounted pump, the first high pressure pump and the first 500-750-1,000-1,250-gpm U. L.-approved pumps. American Fire Pump was very successful in developing new pumping concepts including the only dual shaft series pump, introduced as the DM line, then improved to the only double suction Series H pump line. From 1976 through 1978 American Fire Pump introduced the "IRPS" Series of pumps, which allowed the fire truck manufacturer to mount it as a front mount, mid-ship, power take-off, crash truck, and rear mount application. It proved to be one of the greatest single developments in the fire industry in 50 years, promoting low maintenance, ease of mounting, and service ability. Many of the "IRPS" inter-related pumping systems are still being sold today. They had capacities up to and including 2,500 gpm. Management decided to move the pump operation to a more eco-nomically sound business area and decided on St. Joseph, Tennessee in July 1982. The Company was sold to SPP in 1986, then to Braithwaithe PLC of London, England, then to Hale Pump Company in the 1990s. Mr. and Mrs. Barton retired January 1, 1956, and moved to Sarasota, Florida to a 200-acre ranch to raise cattle as a hobby. On June 24, 1958, the local newspaper, The Stratford Beacon Herald, announced, "Canadian Group Buys Marsh Pumps," coupled with the news of a reorganization for expansion headed by E. W. McIlroy as the President. Late in 1960 American Fire Apparatus sold its minority interest, and its Representa-tives (Directors), Henning O. Anderson and Carleton Leyda, resigned. The plant closed down in 1966.

The second expansion established a manufacturing facility in a 90x181-foot cement and brick building with a pre-cast concrete roof and pillars in 1955 at 801 W. US Highway 30, Marshalltown, Iowa. After the initial contract in June 1954, a lease for 10,000 square feet of floor space was signed in December of that year in a farm machinery plant until a site could be found. The manufacturing plant opened the week of November 19, 1956. Shortly thereafter an addition was completed bringing the total to 90x240 feet (21,600 Square Feet). Dee R. Enos, who was employed by the parent com-pany in Battle Creek for over 19 years, was named the Plant Manager. The gleaming final finish on the trucks was sub-contracted to a facility owned by Ben Niederhauser. During peak production at the Iowa plant they were manufacturing and assembling 25 units monthly and had more than 35 employees. The plant was operational until 1976. Trucks for areas east of the Mississippi River were built at the Battle Creek, Michigan plant and trucks for areas west of the Mis-sissippi were built in Marshalltown, Iowa. Job Numbers were kept separate between the two facilities. While sequenced numerically the same the Marshalltown-built units had the prefix "MC" ahead of the four-digit Job Numbers.

In 1976-1977 American Fire Apparatus entered a sub-contracting agreement with Russell Horse Trailer Company in Bristol, Indiana to produce apparatus, but the agreement had a short life. With all of its manufacturing facilities under way American Fire Apparatus produced over 600 units annually.

In the 1960s, in order to increase sales and enhance the company's image in the marketplace as a full line manufacturer, American Fire Apparatus offered an aerial ladder manufactured by Grove. A three-section 65-foot unit was shipped to Battle Creek, Michigan on January 27, 1960, to be completed and delivered to Honesdale, Pennsylvania. Other aerial units sold over the next 12 years were delivered to Lexington, MD; Lansing, MI; Milan, TN; Lowville, NY (3 S-75 ft); Mayville, NY; Wheat Ridge, CO; New York Mills, NY (4 S-75 ft); Patterson Township, PA; Slippery Rock, PA; Geneseo, IL; Ann Arbor, MI; Tulsa, OK (3 S-75 ft); Laurelton, NY; Wichita, KS; Three Rivers, MI; Plattville, WI; Belmont, MI; North Aurora, IL; and Pine Castle, FL. American Fire Apparatus continued to promote and sell Grove aerial ladder products after the company was acquired by Ladder Towers of Ephrata, PA. Their first unit was delivered to Cicero, IL (4 S-100 ft) with others destined for Oaklawn, IL (3 S-75 ft); St. Petersburg, FL; Fairfax, VA; White Bear Lake, MN; Grand Forks, ND; Fairmont, WV; McPherson, KS; Oshkosh, WI; Battle Creek, MI; Warren, OH; Elkins Park, PA; Park Forest South, IL; Grove City, OH; Clarksburg, WV; Hanover Park, IL; Oakland, FL; Burnsville, MN; Toms River, NJ; Portage, NJ; and the last aerial unit in 1976 was delivered to Orville, OH.

Eager to increase market share, new unique, exclusive products were developed and introduced by American Fire Apparatus. In a 1969 show at the International Fire Chiefs Conference held in Chicago, Illinois was the "Intra-Cab," which provided for their 750-gpm front mounted pump, crankshaft driven, installed behind the custom cab front sheet ahead of the engine. It was installed on an International Harvester custom fire chassis. Also shown was a two-section telescoping water tower-mounted mid-ship, which was later removed and sold as a pumper.

In 1971, at the Chief's convention held in St. Louis, Missouri, a 75-foot 3-section telescoping water tower, named the "Aqua-Jet," which was similar to the Snorkel "Tele-Squrt," was introduced. It was available in lengths of 55 and 75 feet. The unit, mounted on a Ford tilt cab with a 1,000-gpm mid-ship mounted American single-stage pump, was sold to Aurora Park, Illinois. Also in the same year, the Company introduced a custom low-profile 5-man canopy cab, named "The Century 21 Series" (although many of those chassis were Oshkosh, A-Series built) that was delivered to Brooklyn Park, Minnesota. In 1972, the first "Aqua-Jet" 55-foot, rear mounted unit was introduced with a ladder mounted on the top of the telescoping arm, with a nozzle at the end of the boom, all remote controlled from the turntable with a 100-foot cable. The custom chassis unit was delivered to Pine Hills, Florida. In 1971 and 1972 American Fire Apparatus Company added approximately 8,000 square feet of manufacturing space to accommodate the manufacture of the Aqua-Jet line.

The first American/Grove ladder tower, an 85-foot telescoping ladder with platform unit, was shown in 1972, which was mounted on an American custom low-profile tandem chassis and introduced at the International Chiefs Conference held in Cleveland, Ohio. The unit was delivered to Pine Castle, Florida. The years 1974 and 1975 allowed for another new product to be unveiled; a 60-foot 3-section "Aqua-Jet" rear-mounted platform mounted on an American custom single axle chassis with an Intra-Cab and a 1,250-gpm pump. The apparatus was delivered to Rosemount, Minnesota.

Henning J. Anderson, who was the owner and president of the companies, passed away in January 1976 at the age of 81. He is buried with his wife Audrey in a churchyard dating back to 1200 in Markard, Sweden. His two sons, Henning O. and "Andy" W. H. Anderson, took control of the companies.

The fire industry market was changing, and with the emergence of some large New York Stock Exchange Companies now in the marketplace, management decided to sell American Fire Apparatus, which was located at 15150 6-1/2 Mile Road in Battle Creek. American Fire Apparatus ceased operations and closed the facility on February 10, 1978. Faced with orders that couldn't be cancelled, American Fire Apparatus entered into an agreement with Everett Van Wormer, who was their General Manager at that time and had been employed there since 1970, to complete eight contracts. This was the beginning of Wolverine Fire Apparatus Company of Union City, Michigan. Collins Industries Inc. purchased the name and all assets of American Fire Apparatus Company on March 1, 1978. The new Collins Industries-American Fire Apparatus Division was operated in Hutchinson, Kansas where Collins had a 50-acre industrial tract with more than 300,000 square feet of manufacturing facilities along with the corporate headquarters. Collins fire truck manufacturing facilities were completely destroyed by a fire in 1981. A warehouse was remodeled to house fire truck production. Production resumed at a much slower pace. In 1982, they introduced the "Gemini II Series" for commercial apparatus and manufactured their exclusive custom fire chassis, which was identified as the "Golden Eagle Series."

To protect itself from the loss of pump sales of which 50 percent was attributed to American Fire Apparatus sales, a separate subsidiary, PAJ (Plain Jane) Industries of American Fire Pump Company, was established in 1978. The shareholders of American Fire Pump owned the new company. The new company was located in Rogersville, Alabama and offered assembly line-built fire apparatus. The first Plain Jane was actually manufactured by American Fire Apparatus of Battle Creek, Michigan. The unit was introduced that fall at the Chiefs convention held in Cincinnati, Ohio. The unit was constructed on a Ford Model F with a 500-gpm front-mounted pump. It had a 500-gallon tank and was painted Ward LaFrance "lime-yellow" and sold to a dealer for $22,000. Later, 750- and 1,000-gpm units were sold throughout the United States. Several hundred were built and sold within a two-year period until the company announced in June 1981 that they would no longer be involved directly or indirectly in the fire truck business but would concentrate solely on pump and pump accessory sales.

In 1983 the management of Collins Industries felt that the Fire Truck Division, with only two percent market share, was not compatible with the other divisions and the company was sold in 1986 to Harold Locaby of Atlanta, Georgia and relocated to Ball Ground, Georgia. Harold Locaby passed away in December of 1991, and Geraldine, his widow, together with his son, Jerry, operated the company for a few years when Geraldine made the decision to close the facility. The inventory was sold at auction. Jeff Williams of Clay Fire Equipment, located in Ashland, Alabama, and Gerald Shelton of Firemaster Corporation of Springfield, Missouri, both regional builders of equipment, acquired the inventory. This closed the final chapter to one of America's early pioneers in the fire industry, whose stronghold was clearly in producing units on commercial chassis and allowing for the manufacture of some 6,000-plus units since 1936.

However, the final chapter is not written as two different companies have used the American Fire Apparatus Company name early in 2004. We will wait to see what history will dictate.

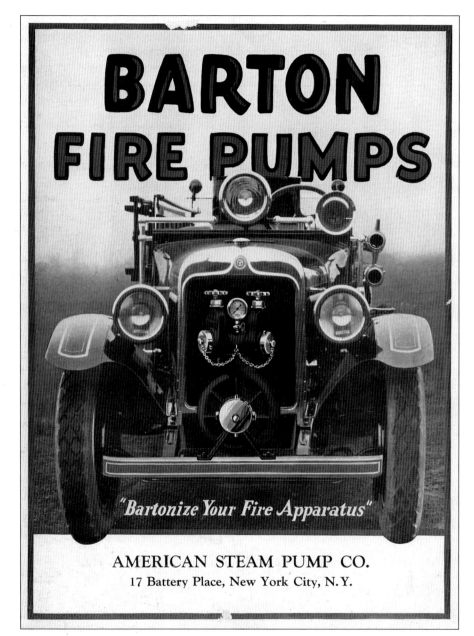

BARTON FIRE PUMPS

"Bartonize Your Fire Apparatus"

AMERICAN STEAM PUMP CO.
17 Battery Place, New York City, N.Y.

Barton Fire Pumps

(Excerpts from the Brochure printed in February 1931)

"CITIES: Every city can fight fire more effectively and with far greater economy by having at least one piece of apparatus equipped with a Barton Centrifugal Underwriters' Fire Pump.

SMALL TOWNS: Disastrous fires have demonstrated time and again that the small town, which relies exclusively upon its water supply system and has no fire pump, is taking a long chance. Water from a hydrant gives neither the volume nor the pressure necessary to quickly and efficiently control a major conflagration. No city would be without pumpers. Why should the small town be less prepared, when a Barton Fire Pump can be had for such a relatively small investment?

RURAL COMMUNITIES: Where there is no water supply system, a Barton pump is the only fire fighting equipment that has real, practical value. It is ideal for villages and rural communities, for fighting fire on the farm, for amusement parks, isolated institutions, etc., because it will pump water from any source. Even when delivering water at suction, this pump develops powerful streams that carry 165 feet and over.

A Barton Fire Pump is a perfect piece of equipment in every way, shape, and manner. There is nothing experimental about it. It is the modern centrifugal type of pump, which means higher efficiency, far greater flexibility, and much longer service. And it is front mounted which means greater economy in both first cost and in upkeep life.

A Barton Fire Pump, no matter how severe the service, will not exhibit any perceptible loss of capacity or pressure, whether it has been on the job five years, ten years, or longer. Although over 400 Barton Fire Pumps are now in service, there has not been a single instance of pump breakage due to a collision. The installation of a heavy bumper will withstand the shock of any ordinary collision and the pump remains intact."

1922 - Early in 1922 a young man named Ben D. Barton had the idea to drive a mechanical device off the front of an engine and convince the management of the American Steam Pump Company of the practical use and its many advantages. Later that year there was several units installed on automobiles owned by contractors used for drainage pumping at construction sites. One afternoon a fire broke out near one of these sites. Someone (it's not known who) got the idea to use this (pumper) to fight the fire with and, thus, the front-mounted fire pump concept had its practical beginnings. The original unit was restored by American Fire Apparatus and used for Public Relations events. A fire destroyed it in 1957 in the Company warehouse where it was housed. *Photo courtesy Keith and Betty (nee Bentley) Mead*

1924 - Mitchfield, Michigan. In the early 1920s, Ford Model T cars and Model TT trucks were the most popular makes in the United States. Many were used in the fire service. In large cities they served as hose wagons and as departmental service trucks. They were the backbones of many suburban and rural fire departments. This picture shows one of the earliest units built with a Barton front-mount pump on a 1924 Ford Model TT truck. *Donald F. Wood & Wayne Sorenson, Big City Fire Trucks, Vol. 1. Photo courtesy Dick Adelman*

1931 - Barton U-34 (300 gpm) Underwriter's Fire Pump Front Mounted on a Chevrolet Six Apparatus. One of the Company's early accommodations came from Max Freeman, Fire Chief of Dundee, Illinois, who wrote, "The Barton Pump has been more than successful in our Department. We have found it always ready to go when the siren gave the alarm. It has given us ample capacities with plenty of pressure under all operating conditions. It is easy to operate, is not injured by dirty water, and does not overheat the motor." Another department wrote, "The Barton Fire Pump continues to prove its reliability in every way." - Fairhaven, Massachusetts. *Photo appeared in Catalog Bulletin No. 60 published in February 1931.*

1931 - Barton U-45 Underwriter's Fire Pump Mounted on a Reo Six Apparatus. B. F. Bracy, Scotland Neck, North Carolina added, "From the roof of an adjoining one-story warehouse, the streams broke the windows on the fifth floor with ease. The pump pressure was 150 to 200 pounds for three and a half hours with two streams running." And Vance F. Barber, Vermontville, Michigan commented, "At several country fires we had to pump very dirty water, which would have been impossible with a rotary gear or piston type pump. We know that the Barton will handle dirty water from a creek, pond, or cistern without any damage to the pump. This fact alone means great savings to any fire department." *Photo appeared in Catalog Bulletin No. 60 published in February 1931*

1931 - Barton U-34 Underwriter's Fire Pump Front Mounted on a Ford AA Apparatus. Other recommendations included, "Shortly after installing the Barton Pump, we had a fire in an eight-room school, the fire having gained great headway before the alarm was given. The truck arrived in record time and by the quick action of the pump, the fire was held to a minimum. Without it the building would have been completely destroyed. We cannot praise the Barton Pump too highly." G. A. Woods, Assistant Chief, New Market, Virginia. *Photo appeared in Catalog Bulletin No. 60 published in February 1931*

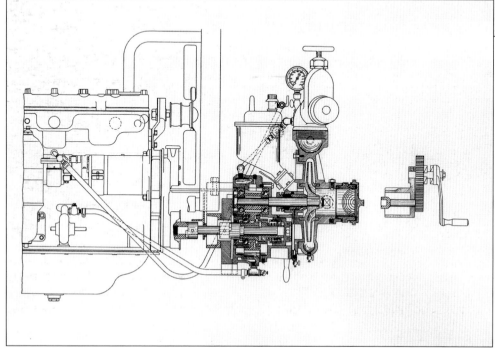

1931 - U-34 Front-Mounted Pump. A sectional drawing showing the working parts of the pump and power take-off which connects pump with motor crankshaft. Since this practical beginning, the Company has been the leader in front-mounted pumps throughout the world, holding the record for the first Underwriters Laboratories front-mounted, high-pressure pump, and the first 500, 750, 1,000 and 1,200 approved front-mounted pumps. In 1972 the Company published that, "between the years 1922 and 1972 it had in service approximately 60,000 front-mount type pump installations in service in every country in the world."

1929 – Manlius, Illinois - An early American Fire Apparatus unit built for the Manlius Fire Protection District. The Chevrolet chassis used here was popular on many early American units. A Barton-American 350-gpm front-mounted pump was specified. An acceptable practice in those times was to have the forward point of the pump ahead of the front bumper, as in this case. A familiar looking tank (200-gallons) was used and mounted behind the cab. The restored unit is in attendance with fire department events and parades. *Photo courtesy of William Friedrich*

1931 - Loma Linda, California - Engine 1130 is a 1931 Ford Model AA with a Seagrave 1916 body re-mounted on the chassis, with a Barton 500-gpm front-mounted pump and a 100-gallon water tank. This unique antique apparatus is used for public relations events and is part of the Department's Fire Museum. The Model AA was chosen after the members of the fire department loaded three different rigs—a Chevrolet, a Dodge, and a Ford. The race was on and the one that made it to the top of Anderson Way (the steepest street in Loma Linda) first would be chosen and the Ford won! The Loma Linda Fire Department originated in 1929. Its origination stemmed from the Loma Linda University Fire Department organized on August 28, 1922. *Photo courtesy of Chuck Madderom*

1932 - Windsor, Pennsylvania - Before a fire apparatus company was formed, this 400-gpm front mount was mounted on a Reo chassis. Little else is known about the cab or the body manufacturer. Note the long sloping front fender, which seemed to be a trend in truck design at that time. The bumper had been extended to accommodate the mounting of the pump. No windshield was provided; a searchlight was mounted on the right-hand cowl side, plus two large red warning lights to alert pedestrians and traffic of an on-coming emergency vehicle. We have no information on the water tank size, but there was a booster reel mounted to the front of the body.
Photo courtesy of Richard M. Adelman

1935 - Durant, Missouri - Another example of the earliest units is this front-mounted pump on a Ford commercial chassis with an open cab. The Barton-American front-mounted pump has a capacity for pumping 500 gpm. The manufacturer of the running board-style body is unknown. Safety has improved as a windshield has been provided, but no wipers. Red emergency lights are provided at the top sill of the windshield, diameter searchlights are mounted at the front right-hand side of the body and a hose reel is evident. Pre-connected soft suction hose is draped from side to side at the front of the vehicle. Wooden ladders are mounted on the left-hand side. *Photo courtesy of Richard M. Adelman*

1936 - West Memphis, Tennessee. Carter Trailer Company located in Memphis, Arkansas, built this rugged looking rig on a Ford chassis. The triple combination pumper utilized a Barton-American 500-gpm front-mounted pump and carried 300 gallons of water. Note the extended front bumper and the unique half-moon racks for the coiled soft suction hose mounted on each side of the pump. A Sterling siren was provided on the crown of the driver's side front fender. A running board-style body was provided. This was the year that American Fire Apparatus was founded. *Photo courtesy of Richard M. Adelman*

1937 - Waterman, Illinois - This model pumper is still the pride of the department at all area parades and public functions supported by the fire department. A 500-gpm front-mounted pump has been installed (note the chassis bumper has not been extended and the pump protrudes ahead of the bumper). A vertical 300-gallon water tank is mounted behind the 3-man coupe cab. A running board-style body with low body side sheet allows for an ample amount of fire hose to be carried. A fabricated hose reel is provided at the forward edge of the body. An electric siren is fender mounted on the left-hand side and a fire bell is fender mounted on the right-hand side. The manufacturer of the body package is unknown. *Photo courtesy of William Friedrich*

1937 - Job No. 208 - Sipesville, Pennsylvania - In 1937 American Fire Apparatus Company was formed. Mr. Barton had the first fire trucks built in Battle Creek in January 1937. Six were built by local body shops. Barton operated the business from his house with Mrs. Barton serving as his secretary. Late in 1937 Barton made an agreement with Anderson Coach Company, manufacturers of house trailers and located 10 miles east of Battle Creek, to build his customer's fire trucks. The first order of 30 trucks built for Sipesville, Pennsylvania was a front-mounted pumper. It is presumed the unit was delivered in early 1938. *Photo courtesy of Bruce Upston*

1939 - Mammoth Cave National Park, Cave City, Pennsylvania - This sleek looking apparatus was manufactured on a Chevrolet conventional coupe cab and chassis. The Barton-American front-mounted 500-gpm pump protruded past the bumper, which was a common acceptable practice in the early years. The booster tank supplied had a capacity of 320-gallons. The only forward red lights were mounted on the front fenders. Note the hard suction hose was carried on the right-hand side. A booster reel was provided at the front of the body. A pipe rack was provided over the body to cover the hose bed with a canvas cover. The apparatus was sold to Muldraugh, Kentucky. *Photo courtesy of Steve Hagy*

1940 - Job No. 236 - Bernie, Indiana - In November 1939 Mr. Barton rented an old brick building on Jackson Street and along with three other men started to build fire apparatus in a 40-foot-square space plus an additional 40-square-foot space for storage. The only equipment they had consisted of a small lathe, a small power hacksaw, an arc welder, one acetylene welder and cutter attachment, several drill motors, a portable sander, and a few other small tools. Sketches were prepared and sent to a local steel supplier for cutting and bending the tank and body parts. The first truck built by American Fire Apparatus in its own plant was a Rural Community truck with a front-mounted pump. It was stationed at Berne, Indiana. *Photo courtesy of Bruce Upston*

1940 - Greensburg, Indiana - This triple combination Ford coupe cab is powered with a 95-horsepower motor and has a Barton-American DMV 500-gpm pump. The cab doors were cutout for the 4-1/2-inch suction that was covered with a wooden housing that became the seat riser. The original rig had two circular hose baskets over the front of the body, which were later replaced with dual hose reels. The water tank capacity is 500 gallons. The compartment ahead of the rear fender contains the truck's radio. The unit retired after 40 years of service. The caretaker, Don Minning, indicates it's currently stored outdoors. *Photo courtesy of Steve Hagy*

1941 - Maple Park & Countryside Fire Protection District, Maple Park, Illinois received their American Fire Apparatus triple combination pumper in August 1941. The fire-fighting package was installed on a Dodge Model WF-32 chassis with a 160-inch wheelbase and a 250-gallon water tank. A Barton U-50 pump was front-mounted and provided 4-inch suction equipment. The total cost of the apparatus on the customer-supplied chassis was $1,916. Extra value features included two Lorraine's lights at the lower corners of the cab, hose reel capacity for 200 feet of 3/4-inch hose, a 30-foot 3-section and a 14-foot roof ladder. One of the day's unique chassis features was a swing-out windshield by Dodge for increased ventilation. *Photo courtesy William Friedrich*

1942 - Rancocas, New Jersey - Truly a credit to the Fire Department as the members of the fire company fabricated and assembled the compartment-style body. Engine No. 2 is a Ford 3-man enclosed coupe designed cab and chassis that features a Barton-American front-mounted 500-gpm pump, and a 500-gallon tank size. Ample compartments were planned for this pumper. Body rails were generously supplied along the top edge of the body and down the sloping beaver tail. All emergency warning lamps were mounted on the top of the roof. *Photo courtesy of Richard M. Adelman*

1946 - Albany, Wisconsin - Here is an example of the quality built into each American Fire Apparatus pumper. It's reported that this apparatus has a 500-gpm pumping capacity. Although the pump is not visible, it is presumed to have a portable unit. The water tank capacity is 250 gallons. Two hose reels are mounted on top of the body. Body handrails have been provided on top of the body and down the rear beaver tails. The apparatus is on duty for fire department functions and is bound to be an eye catcher at parades. *Photo courtesy of William Friedrich*

1947 - Lancaster, Wisconsin - Another parade unit that is a good example of the fire department's care and pride in their equipment. This unique apparatus has both a front-mounted 500-gpm pump and a booster pump. The running board-style body allows for a 500-gallon water tank capacity. The hose reel is transverse mounted for its department's individual needs. Note that the front of the body has been streamlined to the top edge of the conventional 3-man cab. *Photo courtesy of William Friedrich*

1947 - Moose Jaw (bottom) and Wilcox (top), Saskatoon, Canada show just two of the thousands of Willys Jeeps and CJ-3B fire engines made. Many were outfitted with Barton-American pumps ranging from 125-gpm through 500-gpm. The Jeep fire engines didn't receive their own serial number range until 1955. Jeep fire engines were apparently first produced as early as 1945, and a small number were built at the Willys factory in Toledo, Ohio until 1961. Jeeps were also converted by apparatus manufacturers including American Fire Apparatus of Battle Creek, Michigan and Boyer Fire Apparatus of Logansport, Indiana, Central Fire Truck Corp. of St. Louis, Missouri, General Fire Truck Co. of Detroit, Michigan, Howe Fire Apparatus of Anderson, Indiana, John Bean Co. of Lansing, Michigan, Valley Fire Truck of Bay City, Michigan, and W. S. Darley Co. of Chippewa Falls, Wisconsin.
Photos courtesy of Dave Stewardson

1948 - Job No. 618 - St. Mary's, Ohio - The Department took delivery of this classic in 1958. It was built on an International chassis and utilized a Barton-American 750-gpm mid-ship pump and a 500-gallon booster tank. This rig is a good example of the diverse designs used by American Fire Apparatus. Note the roof-mounted Sterling siren light; a stylish looking unit that provided additional rear facing seating for extra firemen. *Photo courtesy of Steve Hagy*

1949 - Job No. 735 - Cadiz, Ohio - Another unique vehicle with a canopy cab design built on a White conventional cab and chassis. The canopy roof is blended on to the cab roof and the body to the rear of the cab. Additional rear seating was provided with access to the rear center aisle-way. A running board-style body was provided. A front fender-mounted siren light is featured. Dual remote-controlled cab spotlights were installed through the cab doorposts. A red flashing beacon ray light was provided. A Barton-American 500-gpm mid-ship pump is supported with a 500-gallon booster tank. *Photo courtesy of Steve Hagy*

1949 - Job No. 936 - Seven Mile, Ohio - This Ford/American Fire Apparatus has a Barton-American 500-gpm pumper with the booster tank removed and a cross-mounted generator installed behind the hose reels. A running board designed body is provided. The Sterling siren light is fender mounted on the left-hand side. A double-faced red warning light is roof mounted. Rear flashing red warning lights are mounted together with swivel hose bed and ground spotlights. *Photo courtesy of Steve Hagy*

1951 - Job No. 1232 - Amelia, Ohio - Not many fire trucks utilize Studebaker chassis, but this community did! This commercial chassis met the Department's needs. A 500-gpm mid-ship Barton–American pump was combined with a 750-gallon water tank. The running board-styled body, overhead ladder rack and dual hose reels all added to its capabilities. The Company expanded its operations in Stratford, Ontario, Canada (1951) to penetrate the Canadian market under the American-Marsh Pumps Canada Ltd. label. The manufacturing facility (Approx. 7,000 square feet) was built at Monteith Avenue in Stratford, Ontario, Canada in early autumn 1951. An additional 5,000 square feet was later added to the plant. During their best years, they produced 150 units a year and had approximately 40 employees. *Photo courtesy of Steve Hagy*

1952 - Job No. 1372 - Kankakee, Illinois - Limestone Township Fire Protection District operated this American Fire Apparatus triple combination pumper. The apparatus is built on a Ford Model F-750 with a 3-man enclosed coupe cab, a Barton-American mid-ship pump with a 500-gpm capacity and a 500-gallon water tank. Running board-style bodies were standard at a time when not many compartment-designed bodies were offered. Hose reels were enclosed with revolving covers. A Sterling siren light was mounted on the forward edge of the cab roof. *Photo courtesy of William Friedrich*

1952 - Job No. 1407 - Plain Township, Canton, Ohio - This handsome "Bulldog" pumper served the fire department for many years until it was sold to Fairfield Township at Summerdale, Ohio. The front bumper was extended for the Barton-American 500-gpm front-mounted pumper. A 1,000-gallon water tank was supplied. Optional extras included the Sterling siren light, red flashing cowl lights, cap remote-controlled cab spotlights, cab beacon ray lamp, dual hose reels, and a rear step personnel windshield plus the standard running board-style body. *Photo courtesy of Steve Hagy*

1953 - Job No. 1460 - Irish Hills, Michigan - One of the most popular fire chassis was the International R Series chassis. This pumper has a Barton-American 500-gpm pump, a 500-gallon tank and a running board-style body. Keeping with the classic American Fire look the rear of the cab is streamlined to the body and ended at the rear cab door line. Covered reels and body handrails were just a few of the optional extras supplied. The unit was two-toned white with green front, rear fenders and wheels. The cab door has a large green shamrock emblem. *Photo courtesy of Steve Hagy*

1954 - Job No. 1658 - Talleyville Fire Company, Wilmington, Delaware - One of the Department's rigs was Engine 251, a 1958 Mack Model L-85 semi-open 3-man cab triple combination pumper. The gleaming fire truck was the pride of its fleet. Beside the chromed grille it had chrome strips on the bumper. Note the chromed rear personnel rear windshield. The Volunteer Fire Company originated in October 1928. In October 2003 Talleyville Fire Company celebrated its 75th Anniversary. *Photo courtesy of Chuck Madderom*

1954 - Derby, Connecticut - Volunteer Fire Department personnel manned this 1954 GMC/American triple combination pumper with a running board-style body. It featured a conventional designed chassis with an enclosed 3-man cab, mid-ship Barton-American cross-mounted 500-gpm pump and a 750-gallon water tank. The front bumper was extended to allow the siren to be center mounted and include space for a fire bell on the left-hand side. Dual boot and coat rails were provided on the top of the body on the left-hand side. *Photo courtesy of Richard M. Adelman*

1954 - Swan River, Manitoba, Canada - This unique extended coupe 2-door cab International R model chassis provided seating for at least five men inside. The pump was a Barton-American 420 Imperial gallon with a tank capacity of 500 Imperial gallons and a running board-style body with a 12-inch kick plate. An overhead ladder rack for the two sections and roof ladder was added. The apparatus is built with the American-Marsh Pumps Canada Ltd. label. *Photo courtesy of Dave Stewardson*

1956 - Job No. 1900 - Greencastle, Pennsylvania - Certainly ahead of its time was this American/International 4-door cab pumper. A mid-ship 750-gpm Barton-American pump was mounted on the International Model R-190 chassis. The back of the cab was streamlined, blending into the front of the pump enclosure, and utilized a 500-gallon water tank and chrome-plated rear personnel windshield. The chassis fuel tank was located under the driver's side crew cab rear door. The bumper was extended for mounting the siren and the top mounted tow hooks. Three strips of the International grille were removed and chromed for an added appearance. Warning and lighting equipment and an electronic siren on the left-hand side were mounted on an extended gravel shield, as well as Unity 6-inch red flashing lamps mounted on the front fenders and rear stanchions, a revolving beacon ray lamp on the cab roof and adjustable white spotlights mounted on the cab cowl and on the rear stanchions. *Photo courtesy of Chuck Madderom*

1956 - Job No. 1923 - Groveland, New York - Churchville Fire Equipment (owned by Jerry Spotts) made its first truck sale on May 4, 1956 to Groveland Fire Department. The front-mounted pumper was manufactured on a 1956 Ford F-800 conventional 3-man cab and chassis with a 175-inch wheelbase and a 25,000-pound GVWR. This unit featured an 18-gallon fuel tank mounted behind the cab seat and a 332 cubic inch 8-cylinder gasoline motor with a 5-speed manual transmission. The pump was a Barton-American UA-50 single stage matched with a 1,000-gallon booster tank. Extra equipment included 4–inch suction equipment. Three lengths were provided on the left-hand side. A single top-mounted booster reel was supplied. The rear side compartment features a sloping rear edge to match the body. Note: 4-digit job numbers without prefix letters were fabricated at the Battle Creek facility. *Factory photo courtesy of Churchville Fire Equipment*

1956 - Job No. 1936 - Bloomfield, Michigan - This certainly falls into the category of one rig you don't see every day. Bloomfield, Michigan was the original owner of this Quint but it was later sold to Somerset Township, Michigan. A White tilt cab was installed on a GMC chassis and Barton-American added a rear-mounted pump. The pump is mounted aft of the rear wheelhouse. A 50-foot Memco aerial ladder was added. The body has high side compartments over the tandem wheelhouse with a larger compartment forward of the body compartments. American built many Quints in the 1950s. *Photo courtesy of Steve Hagy*

1957 - Topeka, Kansas - One of the first units sold to Topeka was this American/International Harvester VCO series tilt cab triple combination pumper—one of the few front-mounted pumps delivered. The bumper has been generously extended to allow cab tilt clearance for the 500-gpm front-mounted pump, which also allowed for a short wheelbase. Water tank size is 500 gallons. Future orders were supplied with extended 6-man canopy cabs that included five units in 1972, a Ford C-Series chassis in 1974, and a 3-man tilt cab and chassis with four-wheel drive in 1975. *Photo courtesy of William Friedrich*

1957 - Carroll Manor Fire Company Inc., Adamstown, Maryland - This department had the distinction of having American Fire Apparatus manufacture a 4-door sedan cab on an International chassis. Coachwork included streamlining the back of the cab and also blending the cab into the pump enclosure; rare options for most manufacturers. A 750 Barton-American pump was mounted mid-ship and the truck carried 500-gallons of water. Optional extra features included a front intake on the driver's side, dual coat and boot rails on the left-hand side, electric hose reel, running board-style body, and an extra length of 4-1/2-inch hard suction hose. *Photo courtesy of Richard M. Adelman*

1958 - Job No. A-222 - Kingsville, Ontario, Canada - Built under American-Marsh Pumps Canada Ltd. is this triple combination pumper on an International Model R-190 series conventional cab and chassis. It features a 625 Imperial gpm mid-ship Barton–American pump with a tank size of 500 Imperial gallons. The apparatus utilized the designs of the parent company in Battle Creek that supplied a compartment-style body. Note the siren was streamlined and mounted and recessed into the forward cab roof. Two booster reels were supplied. On June 24, 1958, the local newspaper, *The Stratford Beacon Herald*, announced a Canadian group purchased Marsh Pumps. The Company was re-organized under E. W. McIlroy who added future expansion. *Photo courtesy of Dave Stewardson*

1960 – Job No. 2311 - Churchville, New York - Churchville Fire Department took delivery of this American Fire Apparatus front-mounted 750-gpm pumper on a Ford F-800 3-man Deluxe cab on September 27, 1960. The 25,000-pound GVWR chassis was on display at the 1960 International Fire Chiefs show in Rochester, New York, September 12 through 15, 1960. A streamlined cab (option "Arcadia" for $225) was selected along with streamlined dual headlamps, plus many other value added features including a 1,000 amp HD Leece Neville alternator system and a 1,200-watt transformer. Pumping equipment included a Barton-American Model UA-75 pump with a 750-gallon capacity, 4-1/2-inch equipment, dual booster reels, and a 500-gallon tank. Body options include sloping tailboard compartments, rear step and personnel windshield. Selling price was $16,100. *Factory photo courtesy of Churchville Fire Equipment*

1960 - Job No. 2338 - Henrietta, New York - The Genesee Valley Fire Department placed their 750-gpm triple combination in service on September 29, 1960. The unique open cab International Harvester R-196 unit was also shown at the IAFC convention in Rochester. It featured a 175-inch wheelbase, power steering, chrome radiator grille bars, headlight rims, parking light rims, front bumper, rear view mirrors, tow hooks, and a Leece Neville 100-amp alternator system. The open cab conversion at that time cost $295. Pumping equipment consisted of a Barton-American Model DME75 two-stage 750-gpm; 4-1/2-inch equipment; two booster reels; and a 750-gallon booster tank. The pump enclosure blended to the cab rear door edge, and it featured extended front and rear compartments with sloping beaver tails and a rear personnel windshield. The contract price with chassis was $15,993. *Factory photo courtesy of Churchville Fire Equipment*

1960 - Job No. M1050 - Eastern Passage, Nova Scotia, Canada - Looking more like its American counterpart is this American-Marsh Pump Canada Ltd. rig built on a Chevrolet 3-man enclosed cab and chassis. The front bumper has been extended for a 500-gpm front-mounted Barton-American pump. A 500-gallon water tank was installed. A compartment-styled body was supplied along with other miscellaneous options including hose reel, warning lights and lighting extras. Late in 1960 American Fire Apparatus sold its minority interest and its Representatives (Directors), Henning O. Anderson and Carleton Leyda, resigned. The plant closed in 1966. *Photo courtesy of Dave Stewardson*

1960 - Bishop, California - The Bishop Fire Department's Engine 4 (top) was the second "Bumper Pumper" purchased from American Fire Apparatus. The triple combination pumper has a 750-gpm front-mounted pump and a 500-gallon water tank. It featured a compartment-style body. The ground ladders were carried in an overhead rack above the hose body. The first unit (bottom) was delivered in 1956 and featured a front-mounted pumper with a 500-gpm pump, a booster tank for 300 gallons of water, and twin hose reels. Both were built on an International conventional 3-man enclosed coupe cab and chassis with the 4x4 four-wheel-drive feature. A standard running board-style body configuration was used on the first unit. *Photos by Chuck Madderom*

1961 - Fairfax County, Gunston, Virginia - Engine 20 was mounted on a Dodge chassis with a 500-gpm front-mounted pump, extended front bumper, and a 500-gallon water tank. This Dodge W-300 featured a coupe cab, a 133-inch wheelbase, a 10,000-pound GVWR Chrysler 318-V gas engine capable of 165-horsepower at 3,900-rpm, and a 4-speed New Process transmission. Body design was standard and functional. Overhead mounted hose reel, and basic warning and lighting equipment were supplied. The Gunston Volunteer Fire Company housed in Station 20 was turned over in January 1974. The Fire Company moved to a new station on May 20, 1976. *Photo courtesy of Chuck Madderom*

1962 - West Adams, Colorado - The county Fire Protection District utilized this two-toned American triple combination pumper to protect the people and property of the district. This unit was built on a 3-man Ford C series tilt cab with a Barton-American 750-gpm mid-ship pump and a 1,000-gallon water tank. It featured lower side body compartments and twin over-the-wheelhouse compartments, extended pump panels, and over-head recessed reels. Top-mounted body handrails finished the body line with a kick-up front panel and chrome plated cab corner brackets for the red emergency lighting. The sloping paint scheme added to the forward motion of the unit. *Photo courtesy of Richard M. Adelman*

1965 - Job No. 2531 - Branchville Volunteer Fire Company, Branchville, Maryland - Few semi-open cab conversions were accomplished. This one deserves to be memorialized. Engine 11, a handsome American Fire Apparatus triple combination pumper built on a Ford C-950 27,000-pound GVWR chassis with a 175-inch wheelbase and a 534-cubic-inch gas motor has well balanced design and proportion. The mid-ship Barton-American pump is a Model DME 1,000-gpm with a 300-gallon tank. Double door compartments are provided forward and aft of the rear wheels. Crosslay hose compartments are provided above the 3/4-height transverse locker compartment. Twin reels are supplied and recess-mounted over the pump enclosure. Originally painted all red, the unit was repainted in the late 1970s to all-white with a lime-yellow band. In 1980 it was sold to PG County for their reserve fleet and repainted red with a white stripe. *Factory photo courtesy of Warren Gleitsmann*

1963 - Job No. 2533 - Alden, New York - The Village of Alden Fire Department took delivery of this 1963 pumper tanker on a Ford Model F-800 on February 21, 1963. Chassis specifications featured a 176-inch wheelbase (103-inch CA), Ford V-8 gas engine that produced 199 horsepower at 3,780 rpm, 25,000-pound GVWR, (front: 7,000 pound axle - rear: 18,500 pound axle), vacuum hydraulic brakes with vacuum booster, 60-amp Leece Neville electrical system, and double channel 21.75-inch section modulus frame. Pumping equipment included Barton-American SSA-1, 500-gpm pump with 4-inch suction equipment, two electric reels for 350 feet of 3/4-inch hose, and a 1,000-gallon tank. Other features were Deluxe running board compartments (28 inches deep), box pan door construction with beveled edges, Morton Kass rear step, hose bed capacity for 1,200 feet of 2 1/2-inch and 600 feet of 1 1/2-inch fire hose. The contract price including chassis was $13,103. *Factory photo courtesy Churchville Fire Equipment*

1963 - Job No. 2545 - Town Line, New York - The Town Line Fire Department placed their front-mounted mini-pumper, one of the few ever made, in service on May 29, 1963. Specifications for the 10,000-pound GVWR Dodge W-300 4x4 are as follows: 133-inch wheelbase; a V-8 engine that produced 165 horsepower at 3,900 rpm; a 4-speed New Process transmission; a Barton-American front-mount Model UA-50 single stage 500-gpm pump; 4-1/2-inch equipment (four 10-foot lengths); one Hannay electric reel; a 200-gallon "Corten" steel booster tank; body compartments standard with rub rails; rear fenderette and sloping beaver tail rear compartment; two seats lengthwise in body from rear of tank to rear of body, with removable seat cushion for access to storage area; also hinged tread plate door under each seat. Selling price at time of bid was $7,290, including chassis and foldaway trailer hitch. *Factory photo courtesy of Churchville Fire Equipment*

1963 - Job No. 2555 - Wolcottsville, New York - Wolcottsville Fire Department accepted their commercial GMC 3-man conventional front-mounted pumper on August 16, 1963. It featured a VH5008 series GMC chassis with a 6-cylinder gas engine; 5-speed manual transmission; 176-inch wheelbase; 102 CA; 27,000-pound GVWR; 7,000-pound front and 18,000-pound rear axles; 9.00 x 20 front and dual 9.00 x 20 rear tires; and a dual battery system of 120 amps. Pump was a Barton-American UA-75 single stage 750-gpm, 4-1/2-inch suction equipment, and had two electric reels, three 10-foot lengths of hard suction hose and an 800-gallon booster tank. It featured a compartment-style body with beaver tail design, plus many other valued options. The contract price with chassis was $14,415. *Factory photo courtesy of Churchville Fire Equipment*

1963 - Job No. 2584 - West Webster, New York - The West Webster Fire Department received this front-mounted pump brush truck on August 16, 1963. The Dodge W-300 4x4 has a wheelbase of 133 inches with a Chrysler 318-V gas motor. The pump is a Barton-American UA-500 capacity 500-gpm with 4-1/2-inch suction equipment and has a square 200-gallon tank mounted in the extreme front of the body, with one left-hand side-mounted booster reel. The body has tread plate kick plates on the body side at running board level, formed 12-gauge tread plate rear fenders, and a tread plate floor in the body behind the tank. Warning and electrical equipment include a Federal #17 beacon ray, two 5-inch rear post-mounted flashers, two 6-inch blue flashing lights on front, a 1,200-watt Leece Neville transformer with 110-volt outlet, and an Edwards cord reel. Contract price including chassis was $6,489. *Factory photo courtesy of Churchville Fire Equipment*

1963 - Job No. 2607 - Hamlin, New York - Hamlin Fire District took delivery of this 750-gpm American Fire Apparatus commercial pumper on January 10, 1964. The chassis is a 1963 Ford F-950, 30,000-pound GVWR 3-man conventional closed cab on a 194-inch wheelbase and 120-inch CA. It is powered by a 534 HD V-8 engine capable of 266 horsepower at 3,200 rpm. It features a 100-amp alternator system and a Barton-American Model DMF-75 multistage 750-gpm mid-ship pump with an extended 1,000-gallon booster tank. Other features include two recessed hose reels with electric rewinds, fully joined cab and body construction without use of a slip joint, four deluxe sides and four rear body compartments, a dual battery system (4-6 volt), two complete 12-volt systems that produce 120 amps per hour, full length dual coat and boot rails, and three 10-foot lengths of 4-1/2-inch suction hose on the left-hand side. Contract price including chassis was $12,958. *Factory photo courtesy of Churchville Fire Equipment*

1964 - Job No. 2610 - Mendon, New York - The Mendon Fire Department placed this canopy cab Ford C-950 750-gpm pumper in service January 10, 1964. Specifications for the 32,000-pound GVWR unit included a 175-inch wheelbase, 148-inch CA, a 534 engine that produced 266 horsepower at 3,200 rpm, power steering, a Spicer 6352-5th direct transmission, a 23,000-pound rear axle, a 100-amp alternator system, and a dual battery system. Canopy cab cost was an additional $1,750. Other features included a Barton-American mid-ship pump Model DME-75, 750-gpm, two AFA electric rewind hose reels, a 1,000-gallon booster pump with removable access covers, two 2-1/2 left-hand discharges, one 2-1/2 right-hand discharge, plus four side and four rear body full through-type compartments. Unit price at delivery with chassis was $23,970. *Factory photo courtesy Churchville Fire Equipment*

1964 - Job No. 2627 - Clarence, New York - The Clarence Fire District accepted the American Fire Apparatus 1,000-gpm triple combination pumper on February 29, 1964. The Ford C-850 Deluxe 3-man tilt custom cab had a 27,000-pound GVWR with a wheelbase of 153 inches, CA of 126 inches, Ford 534 gas engine that produced 266 horsepower at 3,200 rpm, power steering, Spicer 6352 5-speed direct transmission, dual battery system, chrome bumper, grille, tow hooks, and West Coast mirrors, a 9,000-pound front axle and a 22,000-pound rear axle. It featured a Barton-American Model DMF-10 pump, a 1,000-gpm mid-ship pump with standard AFA electric hose reels, and a booster tank capacity of 500 gallons. Chassis features included rear body compartments (full through) with tread plate doors, and four side compartments on each side (one front and one aft of rear axle and two low side compartments above, front and rear.) Delivered price was $20,000. *Factory photo courtesy of Churchville Fire Equipment*

1965 - Job No. 2784 - Darien, New York - The Darien Fire Department responds with their 750-gpm front-mount delivered on August 16, 1965. It is a GMC Model BVH5012, with a COE conventional cab and chassis with a 126-inch CA. It features a Baron-American front-mounted Model UA-75 heated pump with tool compartment (in a stone shield at the right-hand side), a 1,000-gallon booster tank, two electric hose reels, with a capacity for 400 feet of 1-inch hose, a 22-inch locker compartment behind the cab, a 56-inch wide compartment ahead of the rear axle, Morton Kass 20-inch deep rear step, and four compartments at rear of body (two each side of rear center open compartment with two approximately 10-inch high compartments on each side above the lower compartments.) Body and equipment on a customer-supplied chassis was $11,170. *Factory photo courtesy Churchville Fire Equipment*

1964 - North Fork, Indiana - The North Fork Ranger District received this Ford/American Fire Apparatus commercial pumper on an F-Series 3-man conventional chassis. The front bumper is extended to provide for the mounting of a Barton-American 500-gpm pump. The body style selected is a running board-style. The cab has been streamlined to the front of the pump enclosure. The water tank has a 500-gallon capacity. A hinged door is provided on the front right-hand side of the pump enclosure for access to the vehicle batteries. Note the twin Sterling siren lights mounted on the front fenders. It was a true traffic mover! *Photo courtesy of Mark Boatwright*

1965 - Job No. 2802 - Coles District Volunteer Fire & Rescue, Manassas, Virginia - Wagon 8 triple combination pumper chassis is a GMC tilt cab and has a 750-gpm mid-ship pump with a water tank capacity of 1,000-gallons. Note that the rear of the cab has a panel streamlined to the front of the pump enclosure. Other value added features include a covered reel, dual front and rear tool compartments above the side compartments, rear personnel windshield, frame height transverse locker compartment, front cab corner chromed brackets for the siren on the right-hand side and a fire bell on the left-hand side, West Coast mirrors, chrome front tow hooks and a chrome bumper. Warning equipment on the cab includes a beacon ray light, air horn and corner cab spotlights. Four and one half suction equipment was supplied along with two lengths of hard suction hose and a pre-connected soft suction hose. *Photo courtesy of Chuck Madderom*

1965 - Job No. 2809 - Park Hills, Kentucky - A match to a 1959 delivery is this imposing looking open style commercial International V-model cowl and chassis pumper. The classic design provided for a 750-gpm Barton-American pump and a 300-gallon water tank. The compartment style body has a compartment aft of the rear wheel and a wheelhouse compartment above the rear wheel that ended with a rolled edge on the body compartment front sheet. The large, heavy siren was mounted to the cab cowl. *Photo courtesy of Steve Hagy*

1966 - Job No. 2842 - Mayville, New York - The Village of Mayville Fire Department becomes the proud owner of an American Fire Apparatus 65-foot aerial ladder on June 29, 1966. The 1965 IHC-VCO196 tilt cab 35,000-pound GVWR unit has a wheelbase of 153 inches, an FTV549 285-horsepower V-8 engine that produces 3,400 rpm, a dual battery system, 12,000-pound front and 23,000-pound rear axles, a 5-speed transmission, a 100-amp alternator, and power steering. It features a Barton-American 1,000-gpm Model DME-1000 pump with a 300-gallon booster tank, 210 feet of ground ladders, and a Grove 65-foot 3-section hydraulic aerial ladder mounted mid-ship. Honesdale, Pennsylvania received the first American Fire Apparatus Grove ladder truck in 1960. Eight running board compartments were supplied on each side of the rig. Hose bed capacity equaled 1,500 feet of 2-1/2-inch and 600 feet of 1-1/2-inch hose, plus value options. Total apparatus price with chassis was $36,300. *Photo courtesy of Leo E. Duliba*

1967 - Job No. 2927 - Salem, West Virginia - Engine 92, a Ford tilt cab chassis, is equipped with a Barton-American mid-ship 750-gpm pump and a 750-gallon water tank. The 3-man tilt cab was extended to provide for a 6-man canopy cab. Added value features included large flashing cab front sheet-mounted lights, siren and bell with chromed cab corner brackets, dual beacon ray lights, aisle-way compartments, dual recessed hose reels, and top of body handrails. It featured a running board-style body design. *Photo courtesy of Steve Hagy*

1968 - Job No. 3035 - Oregon, Ohio - Here is a long wheelbase Quad unit that provides for ladder storage and numerous compartments. Count them! Two lower, forward of the rear wheels and one aft of the rear wheels, one center above the rear wheel, another upper front of the body, plus the transverse locker compartment behind the 3-man Ford C-Series tilt cab. This Quad seems to have it all: a 1,000-gpm pump, a 500-gallon water tank, and dual hose reels, too. Chrome goodies include a bumper, grille, guide arrow turn signals, red flashing lights, siren, a bell on chrome corner castings, a center-mounted wig-wag light, remote cab spots, roof-mounted air horns and a roof-mounted bar light. *Photo courtesy of Steve Hagy*

1968 - Job No. 3045 - Addison, Missouri - This is a one-of-a-kind International tilt cab series tanker that has a power take-off 250-gpm pump and a 1,000-gallon water tank. A transverse compartment is provided ahead of the tank at the front of the full width body with fenderettes and rub rails. Warning and lighting equipment included a CP-25 siren speaker, cab through the posts spotlights, roof-mounted bar light and rear-mounted swivel hose lamps. *Photo courtesy of Steve Hagy*

1969 - Job No. 3171 - Bergen, New York - The Bergen Fire Department acquired this 2,000-gallon tanker with a front-mounted 750-gpm Barton-American pump on August 21, 1969. The 36,000-pound GVWR Ford Model F-1000 chassis with a wheelbase of 212 inches is powered with a 534 cubic inch gasoline displacement motor. A wide tanker style (not a wet tank style) has four rear compartments with a large center lift-up door. The booster tank is fabricated with #4 gauge (1/4-inch) bottom, #7 gauge (3/16-inch) sides and ends, #10 gauge (1/8-inch) top and baffles, and has magnesium anodes for each 100 gallons of water. An electric reel is provided behind the lift door at the body's rear. Another similar unit (Job No. 2825) was delivered to East Shelby, New York. The 2-1/2-inch hard suction hose is carried on the left-hand side, above the tank, in sliding-type trays, and the 24-foot and 14-foot roof ladders are mounted on the right-hand side. Many other extras were also provided. *Photo courtesy of Churchville Fire Equipment*

1971 - Brooklyn Park, Minnesota - Standard specifications on this Century 21 Series unit include a low profile 5-man cab forward design 84-inches wide with two-piece wraparound tinted safety glass windshield; full view cab instrument panel hinged at bottom, with all necessary gauges, controls and lights; 36,000-pound GVWR rating on all bolted construction, 12,000-pound capacity Rockwell FF-921 front axle, and 23,000-pound Rockwell R-140 single reduction rear axle; full air brake system with 12-cubic foot compressor; rapid build-up Ross power steering; Waukesha-Scania F673DS 6-cylinder 673 cubic inch 320-horsepower turbocharged diesel engine; Fuller HD twin shaft 5-speed constant mesh manual transmission; cooling system with large capacity radiator with exclusive American forced-air intake; and 40-gallon fuel tank, plus a wide variety of options including a black vinyl cab roof. *Photo courtesy of Richard M. Adelman*

71

1972 - Job No. MC-800 - Casselton, North Dakota - The rural fire department provides fire protection for 10 townships: Amenia, Casselton, Durbin, Everest, Gill, Harmony, Maple River, Rush River, Walburg and Wheatland, which cover an area of 410 square miles. The American/Ford C-850 tanker is stationed at Casselton. The 27,500-pound GVWR, 175-inch wheelbase, 148-inch CA (back of cab to center line of rear axle) has a Barton-American Model E1E 750-gpm pump, and a 1,500-gallon water tank housed in a full width body configuration. The tilt cab chassis is powered by a Ford SD-534-4V engine and a Spicer 6352 transmission, and has a Rockwell FE-970, 12,000-pound front axle and an Eaton 18121, 22,000-pound rear axle. Ladders are mounted over the tank, and a transverse locker compartment is provided behind the 3-man tilt cab. In addition to the normal amenities it has a permanently piped deluge gun. The unit was sold in April 2003 to Milnor, North Dakota. *Photo courtesy of Larry Phillips*

1972 - Tappahannock, Virginia - Big, bold and "Bulldog" is this Mack conventional 3-man Model R-Series tandem chassis American Fire Apparatus tanker that carries 1,500-gallons of water. The bumper has been extended to provide protection for the Barton-American 750-gpm front-mounted pump. The compartment-style body has two compartments forward of the rear wheels, two single wheelhouse compartments, and a compartment aft of the rear wheels. Top of body handrails were provided. *Photo courtesy of Richard M. Adelman*

1972 - Job No. 3469 - Orange County, Florida - This Quint served the Orange County Fire Department until 1991 when Panama Fire Apparatus sold it to Winkler, Manitoba, Canada. The original unit was delivered on May 5, 1972. It has a Barton-American 1,250-gpm mid-ship pump and a 200-gallon water tank. An 85-foot Grove ladder tower was rear mounted. The cab was reworked to provide a kick-up roof on each side of the ladder tower, and was enclosed which added two additional doors to the rear canopy. *Photo courtesy of Dave Stewardson*

74

1972 - Richton Park, Illinois - This new product was introduced with various other new product innovations at the International Association of Chiefs Convention held in Cleveland, Ohio in 1972. American Fire Apparatus' new Aqua-Jet AJ 75-foot telescoping water tower, with a vertical reach of 75 feet and a horizontal reach of 63 feet, was mounted on a Ford Model C-Series cab and chassis. The 3-man closed cab was rebuilt to an extended 5-man canopy on a 38,000-pound GVWR chassis, 175-inch wheelbase, and is powered by a Ford 534, V-8 engine that produced 266 horsepower at 3,299 rpm, with a 1,000-gpm mid-ship Barton-American pump. Crosslays open compartments are provided over the front of the pump enclosure. Four outriggers, two forward and two rear of the axle, have a 16-foot spread. The axis of the 1,000-gpm-discharge nozzle is 360 degrees with a nozzle swing of 270 degrees. The overall length is 30 feet with a travel height of 9 feet. *Photo courtesy of William Friedrich*

1972 - Pine Hills, Florida - Also exhibited at the Cleveland convention was American Fire Apparatus' new hydraulically actuated telescopic combination water tower and aerial ladder, named the Aqua-Jet. This Model AJ-55 was 55-feet long and mounted on an Oshkosh Series A custom chassis with a 1,000-gpm mid-ship pump. It featured a low profile 5-man canopy cab and body design with ample body compartments on the lower side and wheelhouse and upper double door compartments, plus a transverse locker compartment with a double crosslay on top. The Aqua-Jet has two outriggers with 12-foot spreads, plus a 55-foot vertical reach and a 43-foot horizontal reach. The turntable has a 360-degree rotation, a 41-inch diameter, a travel height of 8 feet 8 inches, and discharges water at 1,000 gpm. *Photo courtesy of Richard M. Adelman*

1972 - Pine Castle, Florida - Pine Castle Fire Department owned one of the first rigs built with Grove's new aerial tower with a front-mounted basket shown here at the 1972 IAFC Conference in Cleveland. It was built on a tandem, low profile Oshkosh chassis with a mid-ship 1,000-gpm Barton–American pump and an 85-foot Grove ladder tower with platform. In September 1981 Orange County, Florida consolidated 16 fire departments into one. In 1984 the rig was re-powered from a Cummins 903 5-speed Clark overdrive transmission to an Allison automatic transmission and was sold. *Photo courtesy of Richard M. Adelman*

1973 - Job No. 3527 - St Louis, Missouri - McDonnell-Douglas Aircraft has been building aircraft in St. Louis since 1939. Originally known as McDonnell, they merged with Douglas Aircraft of Long Beach, California in 1967. Today, Boeing owns the operation. The facilities for McDonnell-Douglas covered a large area of north St. Louis County near Lambert-St. Louis International Airport. This rig was one of two similar units delivered, the other was built by 3-D in 1978. Built on a Mack DM-685 chassis, this 6x6 unit features a 1,000-gpm pump, a 2,000-gallon water tank, and 210 gallons of AFFF foam. It was designed specifically for fire CFR. The unit is in reserve status today. *Photo courtesy of Dennis Maag*

1974 - Job No. MC-1081/82 - Randolph, New York - The Village of Randolph Fire Department received two 1973 Ford C-900, 31,000-pound GVWR 153-inch wheelbase 3-man tilt cab chassis with cab to axle dimensions of 126 inches on January 23, 1974. Specifications included a Ford V-8 gas engine Model 534, a Spicer 6352 synchromesh transmission, 22,000-pound no-spin rear axle, 12,000-pound front axle, 12-cubic-foot air compressor, and double frame (21.75 section) Modulus. Pumping equipment was an American Model E1G 1,000-gpm mid-ship mounted pump with a 750-gallon water tank. It featured body equipment, locker compartment with full depth compartments and a sloping beaver tail, dual electric reels, 3 10-foot lengths of 5-inch hose, a 24-foot 2-section and a 14-foot roof ladder on the first unit with a 35-foot 3-section and a 14-foot on the second unit, plus many other value added features. Unit cost was $64,987. *Photo courtesy of Churchville Fire Equipment*

1974 - Citrus Heights, California - This unit was originally built as a "Bumper-Pumper" on a Duplex chassis. Engine 23 was rebuilt several times over the years. The front-mounted pump was replaced with a 1,500-gpm mid-ship pump and it carried a 500-gallon water tank. This explains the unique treatment of the cab front. The headlamps were mounted low to eliminate obscuring the light around the front-mounted pump. Citrus Heights Fire District ceased to exist in 1989. The Sacramento County Fire District used the rig for a few years after that and then sold it to a department in Amador County. It was referred to as the "Fat Lady" and, according to Captain Wootton, "was a great rig to drive." Sacramento County eventually merged with another department and formed the Sacramento Metropolitan Fire Department. *Photo courtesy of Bob Allen*

1974 - Fountain Valley, California - Truck 25 was manufactured on a Duplex low-profile chassis, 5-man canopy cab. The 4x6 chassis was required for the 75-foot Aqua-Jet, an exclusive American Fire Apparatus product that served both as an aerial ladder and as an un-manned water tower that featured fog or straight stream with a complete operation from 350 gpm to 1,000 gpm. The mid-ship pump capacity is 1,250-gpm with a 250-gallon water tank. A normal complement, usually 219 feet, of ground ladders was carried. Body compartments included one ahead of the forward extendable jacks, and one forward and one to the rear of the jack on each side. A crosslay compartment is provided above the transverse locker compartment. The rig is painted two-tone Ward LaFrance "lime-yellow" color with a white roof above the cab's beltline and a white ladder. *Photo courtesy of Chuck Madderom*

1974 - Jefferson City, Tennessee - A custom triple combination pumper with a 5-man canopy cab chassis. American's classic design is the sloping rear beaver tail compartments with three overhead body compartments above the standard 166-inch wheelbase body compartments. Other valued options on this 1,250-gpm pumper are an extended bumper with gravel shield, transverse locker compartment, three hose crosslays, dual electric hose reels, plus many more features. *Photo courtesy of Greg Stapleton*

1975 - Job No. 3560 - Lockport, Kentucky - A low profile AFA (probably an Oshkosh Series A) cab forward canopy cab chassis. American had an arrangement with Oshkosh to manufacture a private label chassis for them, the details of which are not known. This rig has an "Insta-Cab" design (with the front-mounted 1,250-gpm pump partially mounted behind the cab front sheet), a 650-gallon water tank and other extras like extended bumper with gravel shield, crosslay compartments, locker compartment, double door high side compartments, dual hose reels and more. After being taken out of service the lime-yellow rig was sold to Kentucky River Fire Department near Lockport, Kentucky. *Photo courtesy of Greg Stapleton*

1975 - Job No. MC-1147 - Labadie, Missouri - The Boles Fire District was formed from the combination of two separate fire departments—Gray Summit and Labadie. Engine 553, a Chevrolet 3-man tilt cab, is equipped with a Barton-American mid-ship 1,250-gpm pump and a 500-gallon booster tank. Other valued features included a transverse locker compartment ahead of the pump and a covered reel compartment. *Photo courtesy of Dennis Maag*

1975 - Job No. 3533 (top) - Chelan County Fire District #1 - Wenatchee, Washington purchased this pair of American Fire Apparatus custom rigs on their own Classic III chassis. Truck No. L-137 has a 1,500-gpm pump, a 200-gallon water tank and a 65-foot telescoping Aqua-Jet 65 water tower and is mounted on a low profile 5-man stationary cab forward model. Note the ground ladders are carried on an overhead rack. Job No. 3534 (bottom) Engine 123 has a 1,500-gpm pump and a 500-gallon water tank and is mounted on a full height stationary cab forward. Both units have high side compartments with separate configurations. Note crosslay compartments and cross-mounted stretcher on top of the hose body behind the recessed hose reel. *Photos courtesy of Bill Hattersley*

1975 - Job No. 3560 - Spanish Lake Fire Protection District, St. Louis, Missouri - The district took delivery of this American "Classic III Series" custom pumper. Built as an "Intra–Cab," the 35,000-pound American Fire Apparatus custom chassis, Model "Classic III Series" has a Detroit Diesel 8V-71N 350-horsepower engine and an Allison HT-740 D automatic transmission, with a 12,000-pound front axle and a 24,000-pound rear axle. The "Intra-Cab" design allowed for a 166-inch wheelbase. The front-mounted single stage Barton-American Model IC-125 1,250-gpm pump was mounted behind the cab front sheet and a 600-gallon water tank was provided. Extra features included double door-high left-hand side compartments, locker compartment, dual crosslay compartments, dual recessed hose reels, and extended front bumper. The rig was sold in 1988 to St. Rose Fire District in Breeze, Illinois. *Photo courtesy of Dennis Maag from the Richard M. Adelman collection*

1975 - Job No. 3625 - Columbus, Ohio - One of four units is this GMC/American 4-door commercial triple combination pumper. The chassis' bumper was extended for the electric front-mounted winch. The enclosed cab featured seating for five firefighters. The narrow painted pump enclosure housed a 1,000-gpm Barton-American pump. It featured a crosslay compartment at the forward edge and a compartment-style body with four vertical high-side compartments and body handrails. The tank held 1,000 gallons of water. *Photo courtesy of Steve Hagy*

1975 - Job No. 3561 (top) - Oak Lawn, Illinois - The Oak Lawn Fire Department received a custom American Fire Apparatus/Oshkosh/LTI 85-foot rear-mounted telescoping ladder tower with platform. The Oshkosh low-profile 5-man cab is on a 228-inch wheelbase and a 4x6 chassis. Truck 12 has a 1,250-gpm pump and carries 300 gallons of water. Four outriggers provided 16 square stabilization areas with four individual controls for four-point leveling on uneven terrain. Job No. 3562 (bottom) is built on an Oshkosh low-profile 5-man canopy cab forward. The 1,250-gpm single stage mid-ship pump (still in service as of this writing) is built on an Oshkosh low-profile cab forward chassis. Pierce Manufacturing Company rebuilt this unit in 1989. *Photos courtesy of Chuck Madderom*

1975 - Job No. 3591 - Garden City, Utah - On September 5, 1975, Garden City received this "all-in-one-piece" American/Ford Model C fire apparatus that ran as Ladder 1. It has a 1,000-gpm pump, a 500-gallon tank and a 55-foot Aqua-Jet telescoping water tower. Full body compartments are provided. Note the irregular design of the compartments that are not consistent with the lower cabinets. Dual crosslays are provided at the top forward edge of the hose bed. Few 55-foot Aqua-Jet telescoping ladder and water towers were built, as the most popular device proved to be the 75-foot Aqua-Jet. *Photo courtesy of Mark Boatwright*

1975 - Rosemount, Minnesota - The Rosemount Fire Department owns this truly unique unit. Built on an American Fire Apparatus custom 5-man canopy cab on a short 166-inch wheelbase. A swing gate has been mounted for added safety for fire personnel riding in the jump seat areas. The Intra-Cab design provides for a 1,000-gpm pump mounted behind the cab front panel. A 300-gallon water tank has been provided. The telescoping and articulating 65-foot platform utilizes many common parts from the exclusive 75-foot Aqua-Jet. It features full body compartments with both lower and high side compartments. *Photo courtesy of Larry Phillips*

1976 - Miamiville, Ohio - This Peterbilt/American caught our attention, as few fire apparatus were built on this chassis at this time. The bumper had to be extended more than normal to allow for clearance between the cab tilt and the 1,000-gpm pump outer housing. A transverse locker compartment was provided behind the cab and a compartment-styled body was supplied. The booster tank has a 1,000-gallon water capacity. *Photo courtesy of Steve Hagy*

1976 - Job No. 3713 - Burnsville, Minnesota - Truck No. 589 was delivered to the Burnsville Fire Department in March 1976. The Quint is built on a Hendrickson 4x6 chassis with a 5-man canopy cab powered with a Detroit Diesel 8V diesel engine and an automatic transmission. It features a Barton-American mid-ship pump with a 1,250-gpm capacity and a 300-gallon water tank. An LTI (Ladder Towers) telescoping aerial 85-foot ladder/platform is mounted on the rear. The original unit was finished in white but now the truck has been repainted red. A normal complement of ladders is carried. *Photo courtesy of Larry Phillips*

1977 - Angola, Indiana - The Angola Fire Department has this version of an "Intra-Cab" in service. Normally the pump is mounted behind the cab front sheet. It's unknown as to why this version was created. The bumper is extended and a housing that shielded the 1,000-gpm pump was externally mounted with the various discharges and suctions, controls and gauges. The chassis is a Spartan "Monarch" series 5-man stationary cab. The tank capacity is 1,000 gallons. Value added features included dual covered reels, transverse locker compartment, and two crosslay hose compartments. *Photo courtesy of William Friedrich*

1979 - Kellnersville, Wisconsin - Another "Plain Jane," Engine 133, on a Ford Model F-700 Series with a 750-gpm pump and a 750-gallon water tank. Three model options were available: "Special" Plain Jane, with 750-gpm front-mounted pump; "Streamline" Plain Jane, with a 1,000-gpm front-mounted pump; and "Super" Plain Jane, with a 1,250-gpm front-mounted pump, high side compartments on the left-hand side (4 doors), 10-foot soft suction hose in lieu of hard suction hose, 2-1/2-inch hydrant inlet, and more. *Photo courtesy of William Friedrich*

1979 - San Bernardo County Fire Department, Adelanto, California - The "Plain Jane" models were offered with limited options and were built on an assembly line. Several hundred were built and sold within a two-year period until the parent company philosophy changed. In June 1981 the company announced they would no longer be involved directly or indirectly in the fire truck business and closed the truck facility but would concentrate solely on pump and pump accessory sales. *Photo courtesy of Chuck Madderom*

1979 - Olive Fire Protection District - Livingston, Illinois - The District purchased the "Plain Jane" pumper (built in Rogersville, Alabama) from Towers Fire Apparatus of Freeburg, Illinois on March 3, 1979, for $30,000. The apparatus was built on a 1978 Ford Model F-700 with a 5-speed transmission and a 2-speed rear axle. It had a Hale 750-gpm front-mounted pump with a 750-gallon water tank. Equipment on the unit included a 35-foot 3-section and a 14-foot roof ladder, a pike pole, a bar light, two cab spotlights, two 4-1/2-inch hard suction hoses, one soft suction hose, a fire axe, a hydrant and spanner wrenches, a siren with a public-address system, and a radio with a microphone. The unit, Engine 301, was sold on June 16, 1996, to Bellflower, Missouri. It had 5,289 miles on it with 538 hours registered on the pump. *Photo courtesy of William Friedrich*

1980 - Eminence, Kentucky - This model still shows the influence of American Fire Apparatus Battle Creek designs. Note the wheelhouse has a semicircle configuration, which was changed in Hutchinson to a flat line on the top of the rear wheelhouse with the introduction of their new body configuration. It features a Barton-American 750-gpm front-mounted pump with an extended bumper and gravel shield. A double door front compartment plus an 18-inch body allows for the 1,000-gallon tank to have a lower silhouette. A locker compartment is ahead of the body with a recessed deck for the dual hose reels. *Photo courtesy of Greg Stapleton*

1981 - Job No. 256 (Note new number system) - Pee Wee Valley, Kentucky - This unit was in the product line of the American Fire Apparatus units when it was a Division of Collins Industries of Hutchinson, Kansas. The product is the "Brush King" Model 300 series. The chassis is a Chevrolet K-20 4x4 with an 8,600-pound GVWR capacity. This conventional 3-man cab and chassis features a 350-horsepower V-8 with a 4-speed manual transmission. It also features a Ramsey 8,000-pound integral mounted with the bumper and the gravel shield. The skid unit that is mounted has a 350-gallon water tank and a 300-gpm pump that is belt-driven from a 10-horsepower Briggs & Stratton air-cooled engine. An electric booster reel is mounted on top of the booster tank. *Photo courtesy of Carl J. "CJ" Haunz*

1982 - Plain City, Ohio - Nothing was plain on this basic GMC "Brigadier"/American Fire Apparatus tanker. This short wheelbase chassis provided 1,000-gpm pumping capacity with the front-mounted pump and roll capability. The bumper was extended and a safety rail allowed for personnel to fight grass fires. A 1,500-gallon tank supplies the water. An extended double door front compartment added to the storage capacity. Two transverse crosslay hose compartments are provided and storage for a portable Fold-a-Tank is available. *Photo courtesy of Steve Hagy*

1982 - Fresno, California - The fire department received this tough looking American/Hendrickson canopy cab forward, stationary 5-man pumper with 1,250-gpm pump and 750-gallon water tank. Body options included both lower and upper high compartments staggered with double door compartments, transverse locker compartment, and crosslay compartment. Value added options are dual CP-25 speakers, dual air horns, a roof-mounted beacon ray light, a remote-controlled left-hand cab spotlight, dual electric reels above the pump, extended front bumper with a storage tray for soft suction hose, and more. American Fire Apparatus made the transition to Hale pumps in June 1981 when they ordered their first Hale pump. *Photo courtesy of Chuck Madderom*

1982 - Job No. F09676FT - Matewan, West Virginia - Engine 5 is a "Gemini Series" apparatus built while it was a Division of Collins Industries. The design provided an independent cross-mounted pumping system with a 1,000-gpm capacity and a standard 750-gallon fiberglass lined tank was supplied. Plenty of storage space is provided with the Gemini modular constructed body offering 110.3 cubic feet of weatherproofing, rust-proofing, and durable vinyl spatter-painted and vented compartments with rear cross-through compartments for even more equipment storage. The apparatus was built on a 27,500-pound GVWR Ford 3-man enclosed tilt cab and chassis. *Photo courtesy of Steve Hagy*

1983 - Sonoma County, Rincon Valley, California - A commercial pumper on a Ford Model C-Series with a 3-man tilt cab. Unit 7585 has a Hale mid-ship 1,000-gpm pump with a 750-gallon tank. Stainless steel pump enclosure panels are standard. Other value added options include high side compartments with lift-up doors, transverse locker compartment, dual crosslay compartments, and a top-mounted electric booster reel. Body side sheets are increased to carry the hose load. The battery box is mounted on the front corner of the running board. The apparatus is finished in lime-yellow. *Photo courtesy of Steve Hagy*

1983 - Job No. 10523 - Yukon, Oklahoma - One of American's "Classic Series" is this front-mounted fire apparatus that was fabricated in Hutchinson. This rugged looking, stubby apparatus has a pumping capability of 750 gpm with the front-mounted pump and extended front bumper. Front mounting the pump offers a point of operation controls and the mounting provides the most serviceable unit, which added convenience for fire department personnel. The GMC conventional chassis has four-wheel drive capability (4x4) and supports a 1,000-gallon water tank. *Photo courtesy of Steve Hagy*

1986 - East Fork, Kentucky - Mack Truck imported the Mid-Liner 28,540-pound GVWR chassis. Mid-Liner powertrain options included either a 175-horsepower or a 210-horsepower engine and a choice of a Spicer 5052A synchromesh transmission, an Allison AT 545 or an MT 643 automatic transmission. Seating capacity is for seven, three in the front cabin and four under the rear canopy. The Mid-Liner features a Hale 1,000-gpm pump with a 1,000-gallon poly tank on an 18-inch body. Three modular high-side compartments are mounted above the standard compartment style body. Dual hose reels are mounted over the pump. *Photo courtesy of Greg Stapleton*

1988 - Fulton County Fire Department - Atlanta, Georgia - Truly a classic, a Mack production Model CF688FC-1327, 5-man cab, (92-inch wide cab) forward design triple combination pumper built with Mack components on a 166-inch wheelbase with an American Fire Apparatus fire-fighting body and equipment. One of the last Mack CF's built is this 1,500-gpm pumper with a 500-gallon poly tank. Optional extras include front suction, extended bumper and gravel shield, two extendable tele-lights mounted in front of the pump enclosure, dual electric hose reels, and high body compartments. *Photo courtesy of Dave Organ*

1988 - Fulton County Fire Department - Atlanta, Georgia - Fulton County's one-of-a-kind Seagrave Model J chassis has a story to tell. Harold Lockaby owned Harold's Sales & Service for years and was a Seagrave and FMC dealer. When FMC decided to build custom rigs, Seagrave gave Harold an ultimatum, and Harold gave up the FMC dealership. Harold needed a commercial apparatus line to sell to the many fire departments that never bought custom rigs, so he purchased American in 1986 and began operations in Georgia. In 1987, De Kalb County ordered a Seagrave tractor to refurb an older Pirsch tiller. Harold's did the refurb with the trailer. De Kalb also had a second Pirsch trailer and Harold, thinking ahead, ordered a second Seagrave tractor. De Kalb never did the second refurb, so Harold built the second tractor into a pumper. It was a demo for a while before Fulton Co. bought it. The unit has a 1,500-gpm Hale pump and a 500-gallon water tank. *Photo courtesy of Dan Decher*

1988 - Vidalia, Georgia - Engine 1 is a Pemfab/American Fire Apparatus "Custom Eagle Series" pumper. The chassis is a "Royale" S-942G 5-man stationary cab, 35,000-pound GVWR chassis powered with a Cummins Diesel Model L10-300 series engine and an Allison automatic MT-647 transmission on a 184-inch wheelbase. It features a 1,500-gpm Hale pump supported by a 500-gallon water tank. The wheelhouse panel is a fiberglass component. Other value added features include extended bumper, stainless steel cab belt trim, stainless steel pump panels, and top-mounted center hose reel. *Photo courtesy Dave Organ*

1989 - Williamson County, Tennessee - This custom rescue apparatus is built on a Spartan "Gladiator" series medium 4-door, 6-man (94-inch wide) tilt cab chassis. Ample compartments are provided on this body. Rescue 12 has the vertical hinged doors. They are double door compartments with each door having its own twist lock. The bumper has been extended to provide for an electric winch. Polished aluminum front fenderettes and truck wheels enhance the appearance of this rig. *Photo courtesy of Greg Stapleton*

1989 - Mallard Creek, North Carolina - This Mack MC cab-over design features outstanding visibility through a two-piece 2,586-square-inch windshield, outstanding maneuverability, plus quick and dependable response. A stationary 2-door cab enclosure was added. It utilizes a Hale 1,500-gpm pump with a stainless steel panel pump enclosure and a poly 1,000-gallon water tank. It also features a compartment-style body with 3 high side compartments with overhead doors, plus many other value added features. *Photo courtesy of Dan Decher*

1990 - Job No. 14726 - Duncan Chapel, North Carolina - A typical view of a Ford tilt cab series with a 175-inch wheelbase pumper. A standard 3-man is purchased and reworked, normally by the fire truck OEM (original equipment manufacturer). Adding a canopy cab extension allows three fire personnel to sit in a protected position facing the rear. Engine 802 has a top-mounted control panel for the Hale 1,250-gpm pump. A 1,000-gallon poly water tank is provided. Under the rear-facing canopy full width seat is a transverse crosslay hose compartment. An 18-inch body has been supplied with three modular high-side compartments with lift-up doors. A deluge gun is mounted above the pump enclosure. *Photo courtesy of Greg Stapleton*

1990 - Jekyll Island, Georgia - Placed into service is this Ford L-800 series/American Fire Apparatus top-mounted control panel triple combination pumper. The 3-man closed cab was extended to a canopy cab and provided seating for three additional fire personnel riding in the rear position. The front bumper is extended, and a stainless steel pump enclosure protects the Hale 1,500-gpm pump. The booster tank is a 500-gallon tank. Other value added features include modular high-side compartments, dual electric hose reel with stainless steel end discs, and a chrome chassis grille. *Photo courtesy of Greg Stapleton*

121

1992 - Hinnesville, Georgia - The fire department accepted this commercial mid-ship Eagle series pumper built on a Ford L-800 3-man conventional closed cab with a Hale 750-gpm pump and a 750-gallon booster tank. The rig has an extended compartment style body. Note the double door compartment ahead of the rear wheels with three modular add-on high side compartments with lift-up door. Extendable tele-lights were furnished on the hose body. Harold Locaby passed away in December 1991 and Geraldine, his widow, together with their son, Jerry, operated the Company for a few years. The Company ceased operations in the summer of 1993. *Photo courtesy of Greg Stapleton*

1992 - Job No. 14816 - College Park, Georgia - This rig was manufactured near the end of production and delivered on March 25, 1993. It went into service as Engine 2 built on a Ford LN-9000, 35,000-pound GVWR chassis with a Cummins 280-horsepower diesel, an Allison HT-740 automatic transmission, a Hale 1,500-gpm Model QG-100 pump, and a 500-gallon poly water tank. The chassis was purchased through State Bid Contract and delivered to American. Extras included an electronic siren recess-mounted in the bumper, a four-door cab with seating for six personnel, an electric hose reel, dual tele-lights mounted to the back of the cab, a portable deluge gun, high side compartments, and more. Contract price was $154,264.92. The apparatus served there until it was auctioned off to Millersville, Missouri and replaced in 1999 by a Pierce "Quantum." *Photo courtesy by Dave Organ*

1993 - Job No. 14845 - Chapel Hill Rural Fire Department, Chapel Hill, Tennessee - One of the last units produced in Ball Ground before operations ceased was this unit. The order was placed on November 2, 1992, and was delivered in July 1993. The apparatus was purchased by the Chapel Hill Lions Club and donated to the Volunteer Fire Department. The "Eagle Series" mid-ship pumper is built on a Freightliner conventional Business Class FL 80 Metro series 35,000-pound GVWR chassis and is powered by a Cummins 300-horsepower diesel engine with an automatic transmission, and a Hale 1,000-gpm pump with a 1,250-gallon booster tank. *Photo courtesy of Chief Paul Rigsby*

Index

Henning J. Anderson, CEO & President

Henning J. Anderson immigrated to the United States from Sweden as a boy of nine, became employed by American-Marsh Pumps Inc. in 1925 and progressed to become the General Manager. By 1936 Mr. Anderson was the President of American Marsh Pumps and then became the Owner and President of both companies shortly after the acquisition of American Fire Apparatus. He passed away in January of 1976 at the age of 81. It was a day like any other as he reported for work, and later in the day, simply put his head in a rest position at his desk and passed away after talking to his wife at 4:30, asking her if she had everything she needed. When she asked him why he was talking that way, he said, "I'm not going to be here tomorrow." He is buried with his wife Audrey in a church yard dating back to 1200 A.D. in Markard, Sweden.

About the Author

Richard J. Gergel

American Fire Apparatus Company deserves to be memorialized as throughout its history, it has manufactured every type of fire apparatus you can imagine. I was associated with the Fire Industry for nearly 45 years and would like to "leave a little something behind." This is my 3rd photo archive book, co-editing *Ward La France Fire Trucks 1918-1978 Photo Archive* and then authoring *Imperial Fire Apparatus 1969-1976 Photo Archive*.

My first employer was Ward La France, Elmira Heights, NY, where I served from 1951 through 1979, in various positions from blue print operator to the Company's President. Next I joined Mack Trucks in Allentown, PA as the Fire Apparatus Division Sales Manager. Four years later [1985], I changed employment opportunities with Pemfab Trucks in Rancocas, NJ, where I retired on December 31, 1995.

I recently received a letter from Gary Handwerk that I'd like to share an excerpt with you: "As the last chief engineer of American Fire Pump, I had access to many of the old records, dating back to the '20s, from both the pump and apparatus companies. What I saw was a company who produced a remarkable number of apparatus and who was, for many years, an innovator and major force in the industry. In the mid '80s many of these records we lost, But your research has uncovered more information than even I believed still existed. For many of us who have a passion for fire apparatus, books like this bring back many memories and makes us realize that without people, like you, who take the considerable time and energy to record this history, it would be lost forever. I look forward to your next project."

During the compilation of this photo archive, I compiled a registry of over 2,500 units built. It's important to add and maintain this registry; please email RJGergel@aol.com or write to Richard Gergel, C/O SPAAMFAA, PO Box 2005, Syracuse, NY 133220 or write to Iconografix indicating you would like your letter (with proper postage provided) forwarded. Please indicate job number, date of manufacture, current owner, model, and your address.

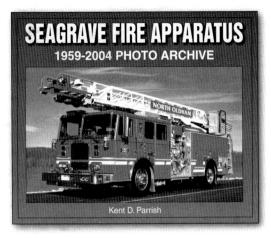

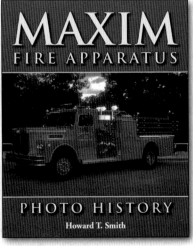
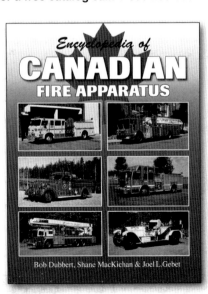